Health 239

母乳的故事

Mother's Milk

Gunter Pauli

[比] 冈特·鲍利 著

[哥伦] 凯瑟琳娜·巴赫 绘

章里西 译

上海远东出版社

丛书编委会

主 任: 贾 峰

副主任: 何家振　闫世东　郑立明

委 员: 李原原　祝真旭　牛玲娟　梁雅丽　任泽林

　　　　王 岢　陈 卫　郑循如　吴建民　彭 勇

　　　　王梦雨　戴 虹　靳增江　孟 蝶　崔晓晓

特别感谢以下热心人士对童书工作的支持:

匡志强　方 芳　宋小华　解 东　厉 云　李 婧

刘 丹　熊彩虹　罗淑怡　旷 婉　杨 荣　刘学振

何圣霖　王必斗　潘林平　熊志强　廖清州　谭燕宁

王 征　白 纯　张林霞　寿颖慧　罗 佳　傅 俊

胡海朋　白永喆　韦小宏　李 杰　欧 亮

目录

Contents

从北极回来的路上，一头小座头鲸在妈妈旁边游泳。她可能很快就要离开妈妈独立生活了，所以她转向妈妈说：

"妈妈，非常感谢您对我的照顾。但现在是我该对自己负责的时候了。天晓得，也许有一天我也会有一头自己的小座头鲸。"

"你不再需要我了，我感到非常骄傲。你已长成一个智慧的姑娘了。"

A humpback whale calf is swimming next to her mother on the way back from the Arctic. She may soon continue life without her mother, so she turns to her and says:

"Mom, I am so grateful to you for taking such good care of me. But the time has come for me to be responsible for myself. Who knows, one day I may even have a calf of my own."

"I am very proud that you won't be needing me anymore. And that you have grown up to be a wise young lady."

一头小座头鲸……

A humpback whale calf ...

还有你爸爸……

And if it weren't for your dad ...

"哦，如果不是因为您，我就不会在这里了。"小座头鲸回答。

"当然，还有你爸爸。我知道他也很在乎你。"

"我一直在想，您是如何给我提供了那么多乳汁……几百升几百升的，快一年了，每天都有。"

"Oh, if it wasn't for you, I wouldn't be here," the calf replies.

"And if it weren't for your dad, of course. I know he also cares a great deal."

"I have always wondered how you were able to provide me with so much milk… hundreds and hundreds of litres, every day for almost a year."

"嗯，鲸做的每件事都很大。我们每一次心跳都会泵出1 000升血液，还有，我为你产了好几吨的乳汁，这样你才能够健康成长。"

"这真是个奇迹，妈妈。我每天都吃奶，而且常常是在你自己一整天都没吃东西的时候吃。"

"Well, everything a whale does is big. We pump a thousand litres of blood with every heartbeat, and I produced tons of milk for you, so you could grow, and be healthy."

"It is such a miracle, Mom. All that milk day after day, and often when you yourself have not eaten at all that day."

每一次心跳都会泵出1 000升……

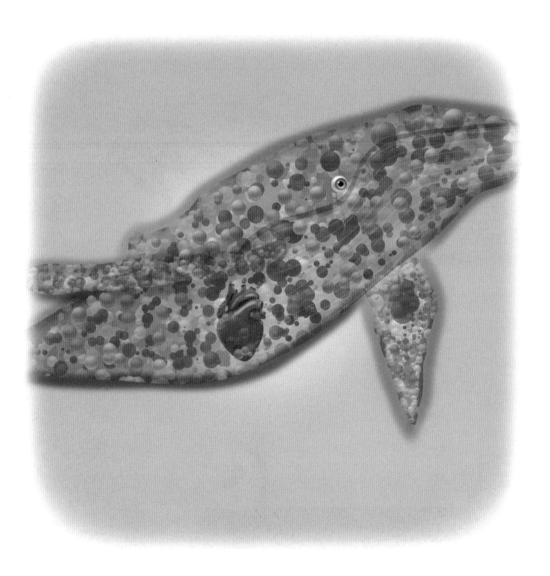

a thousand litres with every heartbeat ...

"亲爱的，你可能很想知道母乳到底有多少，但是母乳所带来的能让你健康度过一生的那些好东西，你都知道吗？"

"哦，当然知道。我知道有糖和脂肪，以及那些构建我的骨骼的矿物质。"

"You may well wonder about the volume of the milk, my dear, but what about all the good things that come with it, that makes you healthy for the rest of your life? Are you aware of those?"

"Oh yes, I am. I do know about the sugars and the fats, and the minerals that build my bones."

"很好，但可能同样重要的是，我的乳汁喂饱了你的肠子，它也喂饱了肠道里所有的虫子、真菌和病毒，"鲸妈妈解释说。

"我的肠道里到处都是细菌，甚至还有很多真菌和病毒。我明白这些微生物会让我变得更强壮，适应力更强。它们甚至还为我的大脑提供充足的食物。"

"那住在你肚子里的那些小虫子是谁给你的？"

"Good, but what may be as important is that my milk fed your gut, and it fed all the bugs, fungi and viruses in your gut too," her mother explains.

"My gut is full of bacteria, and even rich in fungi, and viruses. I realise these microorganisms make me strong, make me resilient. And that they even feed my brain."

"And who gave you all these little bugs that now live in your tummy?"

……它也喂饱了肠道里所有的虫子、真菌和病毒……

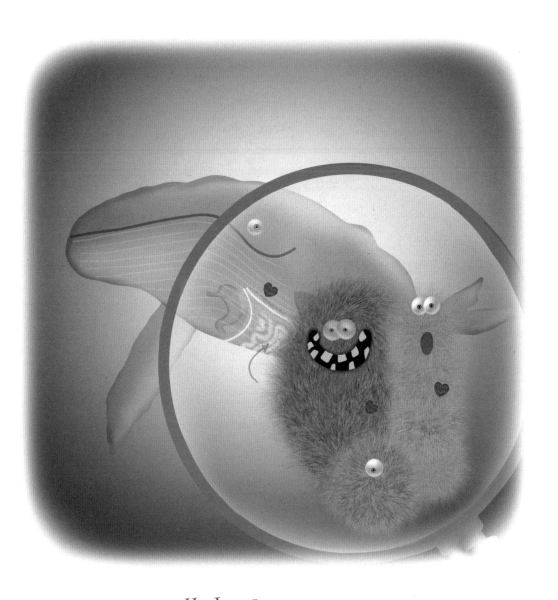

... and it fed all the bugs, fungi and viruses ...

在您的子宫里……

In your womb …

"我不知道，它们可能只是某一天出现的……"

"告诉我，在你出生并开始游泳之前，你在哪里生活和成长?"

"当然是在您的子宫里。我什么都不记得了，不过既然您问了，我猜想它们很可能是从您那儿弄来的! 当我在您的身体里，还有您生我的时候。"

"I don't know, they probably just came along one day... "
"Tell me, where were you living and growing before you were born and started swimming?"
"In your womb, of course. I don't remember anything about that, but now that you're asking… I probably got them from you! While I was inside you, and on the way out."

"大自然多么精彩！每一种新生物都需要有自己的完美微生物组合，所以妈妈们提供的食物中还包括益生菌生长所需的一切。"

"有意思，您把您的一些细菌传给了我，它们在我的口腔和肠道里生活。我们需要快乐细菌的混合组来一起工作。"

"How brilliant is Nature! As every new creature needs its own good mix of microbes, the food their mommies provide includes everything the good bugs need to thrive."

"Now that's interesting, that you passed some of your bacteria on, to come and live in my mouth and my gut. We need a mix of happy germs to work together."

......快乐细菌的混合组来一起工作。

... a mix of happy germs to work together.

……太感激您了，妈妈……

... I am so grateful, Mom ...

"我的乳汁里几百种糖的特殊混合物吸引了数万亿健康的虫子，它们将伴随你的一生。"

"有一天，我也会把它们传给我的小座头鲸们！太感激您了，妈妈。您不仅给了我生命，用乳汁哺育我，还给了我许多保证我健康的小朋友们。"

"The special mix of hundreds of sugars in my milk attracts trillions of healthy bugs, and these will now accompany you – for the rest of your life."

"And, one day, I will pass them on to my calves! I am so grateful, Mom. You not only gave me life, and gave me milk, you also gave me lots of tiny friends that will ensure my health."

"我并不是唯一传给你虫子的人。每次爸爸带你游泳，和你一起玩耍时，他都在丰富着细菌的组合。你的健康是爸爸和我，还有祖父母共同给你的。"

"感谢你们给予我健康。在我离开之前，让我再拥抱您一次，来表达我的爱和感激……同时，也为了在路上捉到好虫子。"

……这仅仅是开始！……

"And I am not the only one who donated them. Every time your dad swam and played with you, he too enriched the mix of bugs. This is how you are the product of you dad and myself, and your grandparents too."

"Thanks for my health. Before I leave, let me embrace you once more, to show my love and appreciation … and to get a great shot of good bugs for the road."

... AND IT HAS ONLY JUST BEGUN!...

······这仅仅是开始! ······

... AND IT HAS ONLY JUST BEGUN! ...

在婴儿出生的最初几天，数以百万计的细菌寄生在婴儿身体上，附着在婴儿的皮肤、口腔，尤其是肠道里。这些细菌通过产道从母亲转移到婴儿身上，也可以通过皮肤接触转移。

In the first few days of a baby's life, millions of bacteria colonise the baby's body and settle on its skin, in its mouth, and especially in its gut. These are transferred from mother to baby in the birth canal, and are also transferred through skin-to-skin contact.

这些寄生微生物对婴儿的健康有着终生深远的影响。通过剖腹产出生的孩子接触这些物质较少，而且似乎有更高的肥胖、患糖尿病和哮喘的风险。

The colonising microbes have a far-reaching, life-long impact on the baby's health. Children born by caesarean section are less exposed to these, and seem to have a higher risk of obesity, developing diabetes and asthma.

Breast milk, along with the bacteria that colonise the baby's intestines, secures the child's growing immune system, as well as the baby's digestion. Breast milk itself is rich in beneficial bacteria.

母乳和寄生在婴儿肠道的细菌一起，保护了婴儿正在发育的免疫系统以及消化系统。母乳本身就富含益生菌。

Breastmilk contains up to 200 different ingredients, some which attract specific bacteria. Many of these ingredients are not present in cow's milk. Antibiotics taken by mother or baby obliterates the nascent bacterial population.

母乳含有多达 200 种不同的成分，其中一些会吸引特定的细菌。许多这些成分不存在于牛奶中。母亲或婴儿服用抗生素会杀死新生细菌。

One family of beneficial microorganisms is the bifidobacteria, which sense and respond to hormones. These bacteria consume special and complex sugars that babies cannot digest without them. None is available in formulated infant food.

一种有益微生物是双歧杆菌，它们能感知激素并做出反应。这些细菌消耗特殊而复杂的糖，没有它们婴儿就无法消化。婴儿配方食品中不含这些成分。

The benefits of breast-feeding, and a mother's loving contact, are that these enrich the baby's body with microbiomes essential for its health. These are transferred to the baby's mouth, skin and gut in colonies that keep on evolving.

母乳喂养和母子亲密接触的好处是丰富对婴儿健康至关重要的微生物。这些微生物被转移到婴儿的口腔、皮肤和肠道，并不断进化。

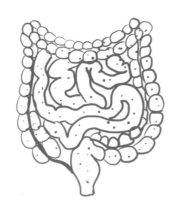

The first 1000 days of life lay the foundation for a child's future health. During this critical period, the gastrointestinal tract undergoes rapid and evolving colonisation by microorganisms, including an estimated 100 trillion bacteria.

生命最初的 1000 天为儿童未来的健康奠定了基础。在这一关键时期，胃肠道经历了微生物的快速进化，其中估计有 100 万亿个细菌。

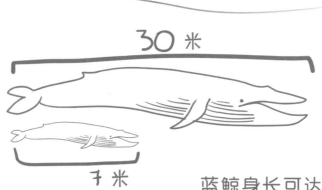

Blue whales reach 30 metres in length, weighing up to 150 tons. They are found in every ocean. Females give birth every 2-3 years. Calves are 7 metres long, weighing 2.5 tons, and require 200 litres of milk to increase body weight by about 100 kg per day.

蓝鲸身长可达 30 米，重达 150 吨。在每片海洋中都能找到它们。雌鲸每 2 到 3 年生产一次。小鲸身长 7 米，重达 2.5 吨，每天要吃 200 升奶才能增加约 100 千克体重。

Many bacteria are good for you, agree?

很多细菌对你是有好处的，同意吗？

Playing is fun, but do you like the idea that this gives you more bugs?

游戏很有趣，但是你喜欢会给你带来很多小虫的玩法吗？

What does your gut have to do with your brain?

你的内脏和你的大脑有什么关系？

Is it a miracle that there are bacteria everywhere, or is this Nature's design?

细菌无处不在，这是一个奇迹，还是大自然的设计？

It is a fact that we live in symbiosis with bacteria, but do we know where all the bacteria that inhabit our mouth, skin, and intestines come from? Ask around to see if your friends and family members know the origin of these bacteria. First listen to what they have to say, and then share the information you have learnt from this fable, and the insights you have gained. Now, there is a second question we need to ask: how can we ensure that we continue to have the bacteria that are good for us, while not allowing the bacteria that are bad for us, to thrive? List the steps we can take to ensure that only the good bacteria thrive. Share these with your friends.

我们与细菌共生，但是你知道生活在口腔、皮肤和肠道中的细菌是从哪里来的吗？问问身边的亲朋好友。先听听他们怎么说，然后分享你从这个寓言了解到的信息，以及你的感想。现在，回答第二个问题：如何确保我们长久拥有益生菌，同时又不让有害菌泛滥？列出我们可以采取的步骤，确保只有益生菌茁壮成长。和朋友们分享这些信息。

学科知识
Academic Knowledge

生物学	有益双歧杆菌生长较快；座头鲸每年只有120天在进食；下丘脑腺垂体控制催乳素；甲状腺在激素产生中的作用；幼鲸消耗的乳汁占其体重的2%到10%；鲸一次可以在胃里储存一吨的食物；母乳是个性化的；最初的母乳(初乳)覆盖在胃肠道上，奠定了后代免疫系统的基础。
化 学	母乳中的糖是低聚糖；母乳喂养所必需的催产素和催乳素；雌性激素会增加催乳素的水平；母乳含有200种不同的成分，包括蛋白质、脂类、碳水化合物、抗体、激素、酶、维生素和矿物质；肠用免疫球蛋白IgA有通便作用，排除不良物质。
物 理	生理刺激乳头和乳房会释放催乳素；催产素引起泌乳反射；鲸乳脂肪含量高，在水中看起来像牙膏。
工程学	牛生长激素是通过基因工程来获得的，可刺激奶牛产奶。
经济学	自然分娩和母乳喂养为社会节省的保健费用；剖腹产的费用以及由于没有机会从母亲那里获得个性化母乳而对健康造成的后续影响；对社会而言，治疗胃肠疾病的费用高于治疗心脏病、外伤的费用。
伦理学	孩子出生后给予父母产假的公司；个性化母乳和世界标准化配方奶的区别；过量使用抗生素，影响肠道菌群的健康，尤其是儿童。
历 史	欧洲的皇室用奶妈来给婴儿喂奶；捕鲸船从16世纪开始在北极捕鲸。
地 理	北大西洋和北太平洋有3万头座头鲸，每头鲸每年迁徙2.5万千米；欧盟禁止使用牛生长激素。
数 学	关于如何扭转日本、韩国、意大利和德国人口不断减少的数学模型：通过模型显示，母乳喂养每年需要1 800小时（每天5小时，每周35小时，包括周末）。
生活方式	努力工作的价值和重要性体现在母亲将母亲身份与所有其他活动结合起来；拥抱、关心、照顾、耐心、接受孩子本来的样子，当然还有感激的重要性；母鲸和幼鲸一生的关系。
社会学	越来越少的父母花时间和孩子在一起；工业化国家迅速下降的出生率；用罐装奶喂养的虎鲸和海豚寿命较短，因为持续圈养导致生存压力水平较高。
心理学	情感依附的第一个纽带是与母亲的关系，通过母乳喂养形成与日常生活息息相关的信任、安全和敏感的基础；父亲教导孩子如何对待母亲，强化恰当的地位和自尊；催产素是母子关系的关键。
系统论	自然分娩的婴儿出生时是无菌的，母亲和婴儿之间的皮肤接触是必不可少的，与父亲和亲属的接触也是如此；至少一个星期不要给宝宝洗澡，因为宝宝皮肤上的液体是独特的菌群来源。

情感智慧
Emotional Intelligence

鲸女儿

小座头鲸很感激她的妈妈。她自信现在可以独立生活。小座头鲸感谢妈妈给予她生命，在受到启发和激励后，她梦想未来自己也会创造新的生命。她自信地询问一些核心问题。她认为妈妈能产那么多的乳汁是个奇迹。虽然小座头鲸知道母乳的种种好处，但当听说肠道菌群是从母亲、父亲和亲属那里转移过来时，她还是很惊讶。她承认自己对有益菌群是如何转移和保持缺乏理解，这些信息会让她拥有积极热情的态度，有机会过上健康的生活。

鲸妈妈

鲸妈妈为成功将女儿抚养成人并独立生活而感到自豪。她呼吁人们认识到父亲的重要性。关于大小，鲸妈妈解释说鲸很大，所以所做的每件事都很大。鲸妈妈希望女儿更多地了解健康是如何依赖于婴儿期获得的母乳。鲸妈妈说出了令女儿惊讶的信息，并激励她去猜测体内菌群是如何建立起来的。鲸妈妈让女儿的认识更进了一步，即益生菌会陪伴她的女儿，在她的一生中不断进化，从而更有利于她的健康。

艺术
The Arts

用牛奶来做一个有趣的艺术活动吧。你需要一个碗，一杯牛奶，一些皂液，一个棉签，食用色素和一些胡椒粉。把牛奶倒进碗里，然后小心地在牛奶表面滴几滴食用色素。用棉签蘸取皂液，然后轻触牛奶表面的色素小滴。你看到了什么？重复这个过程，但在用皂液棉签轻触之前，在牛奶上撒一小撮胡椒粉。发生了什么？观察仅用天然原料的创作过程和白色背景下奇妙的色彩变化。

思维拓展
Systems: Making the Connections

人体是各种细菌、古细菌、病毒和真核微生物的支架。它们栖息在人体小生境中，数量比人体细胞多一个数量级。总的来说，胃肠道微生物联盟相当于"一个器官中的虚拟器官"。对婴儿来说，适宜的细菌、真菌甚至病毒都是通过与母亲的皮肤接触，以及与父亲和直系亲属的拥抱照顾获得的。所有的接触、触摸或呼吸，促使益生菌种群在肠道、口腔和皮肤养成。母乳中有200多种不同的成分为宝宝量身定制。母乳是关键，它优于配方奶的好处已得到充分证明。这也许是我们都应该学习的共生和互惠关系中最重要的一课。对所有哺乳动物来说，这种与微生物的关系是健康生活的关键。我们身体的耐受力依赖于微生物。因此，对抗生素和抗病毒药物的依赖和过度使用，降低了我们免疫系统自我保护的能力。按照同样的逻辑，一个生态系统的恢复力依赖于所有物种的活力。只有当我们明白了创造有利环境来欢迎益生菌的重要性，明白了同样的免疫系统是如何抵御有害菌的，我们才能相信大自然对生命的设计。紧密的母婴关系清楚地展示了这一点。

动手能力
Capacity to Implement

为了让所有人相信母乳对下一代健康很重要，我们来看看它的好处：母乳喂养对母亲和孩子的健康都有益，即使孩子早已过了婴儿期。这些好处包括适当的热量产生和脂肪组织发育，婴儿猝死综合征的风险降低73%，提高智商，降低耳部感染，提高对感冒和流感的抵抗力，略微降低儿童患白血病，降低儿童期糖尿病风险，降低患哮喘和湿疹的风险，减少牙齿健康问题，降低晚年肥胖的风险，降低患心理障碍的风险。请宣传这些信息，激励所有对母乳喂养抱有疑问的未来父母们。

故事灵感来自
This Fable Is Inspired by

梅根·阿扎德
Meghan Azad

梅根·阿扎德生于加拿大。2004 年她在温尼伯大学获得生物化学文凭。2010 年，她在曼尼托巴大学获得生物化学和医学遗传学博士学位。2015 年，她在阿尔伯塔大学获得班廷博士后奖学金，重点研究儿科。2018 年，梅根在伦敦卫生与热带医学院获得流行病学理学硕士学位。她是儿科和儿童健康方面的助理教授，也是儿童医院研究所的研究员。她对婴儿的健康发育很感兴趣，致力于研究婴儿的发育起源或慢性疾病。她是国际母乳和哺乳研究协会的秘书。目前她的研究项目侧重于产妇营养、婴儿喂养和母乳构成。

图书在版编目（CIP）数据

冈特生态童书.第七辑：全36册：汉英对照 /
（比）冈特·鲍利著；（哥伦）凯瑟琳娜·巴赫绘；
何家振等译. —上海：上海远东出版社，2020
ISBN 978-7-5476-1671-0

Ⅰ.①冈… Ⅱ.①冈… ②凯… ③何… Ⅲ.①生态
环境 – 环境保护 – 儿童读物—汉英 Ⅳ.①X171.1-49

中国版本图书馆CIP数据核字（2020）第236911号

策　　划	张　蓉
责任编辑	程云琦
封面设计	魏　来 李　廉

冈特生态童书

母乳的故事

[比]冈特·鲍利　著

[哥伦]凯瑟琳娜·巴赫　绘

章里西　译

记得要和身边的小朋友分享环保知识哦！
八喜冰淇淋祝你成为环保小使者！